Hands-On Projects

THE FARM

Active Learning about Plants and Animals

by Carol Wawrychuk & Cherie McSweeney
illustrated by Philip Chalk

Contents

For a complete catalog, please write to the address below:
P.O. Box 1680, Palo Alto, CA 94302

ISBN 1-57612-013-9
98765432

Introduction

How many times have you heard a child ask, "Where does milk come from?" or "How did that plant grow?" Have you ever wished a child could experience life on a farm?

The Farm is a thematic unit that allows children to explore a country setting and farm environment. If it's not possible to take the children to a real farm for a field trip, this unit can bring an imaginative farm to them!

Large boxes become a barn, tractor, and fruit and vegetable stand. Imaginative play comes alive as the children dress up in costumes they have made. Some children play at being pigs, ducks, or horses living in the barn. Others wear their farmer's hats and vests and "plow" the fields in the tractor. Still more "sell" produce at the fruit and vegetable stand. And all children love the experience of "milking" the sawhorse cow!

The process of seed growth is realized as children plant and care for their own seeds in bottles. This activity is reinforced through a group mural and individual projects.

Personal Observations:

The children eagerly used the dramatic play props in their imaginative farm environment. Some of the children made a barnyard from blocks in which they pranced around on their riding ponies. Others pretended they were pigs at a feeding trough. Quacking sounds could be heard as children helped one another put on their duck headbands and feet.

The boys and girls loved making and wearing the farmer's hats and vests. They wore their farm outfits as they used a wooden wheelbarrow and buckets turned upside down for stools. Milking the cow was a most popular activity. The latex glove constantly needed refilling as the children squirted milk into the bucket.

Children experimented with weights and measures as they filled the shovel of the tractor to see how much it lifted. They discovered that if several children worked together, more dirt could be lifted. Laughter rang out as the items tumbled from the shovel into the tractor.

It was a delight to stand back and observe the children engage in cooperative play as they re-enacted the farm experiences. It truly became their imaginary world!

Big Red Barn

Materials:
Two large appliance boxes, heavy string or yarn, tempera paint (red, brown, and white), paintbrushes or rollers, shallow tins (for paint), sharp instrument for cutting (for adult use only), scissors

Directions:
1. Remove one end of one box (making sure the other end is left intact).
2. Cut a pitched-roof shape on the front and back of the box.
3. Cut a T-shaped door in the front of the box and a flap for the loft.

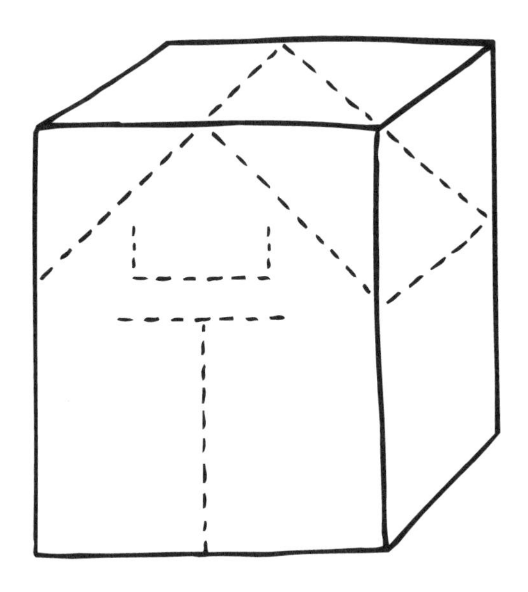

Big Red Barn

4. Cut a section out of the second appliance box for the roof, and fold it in half.
5. Punch and align holes on the roof and the pitched roof-shape box, and attach the barn and roof with string or yarn.

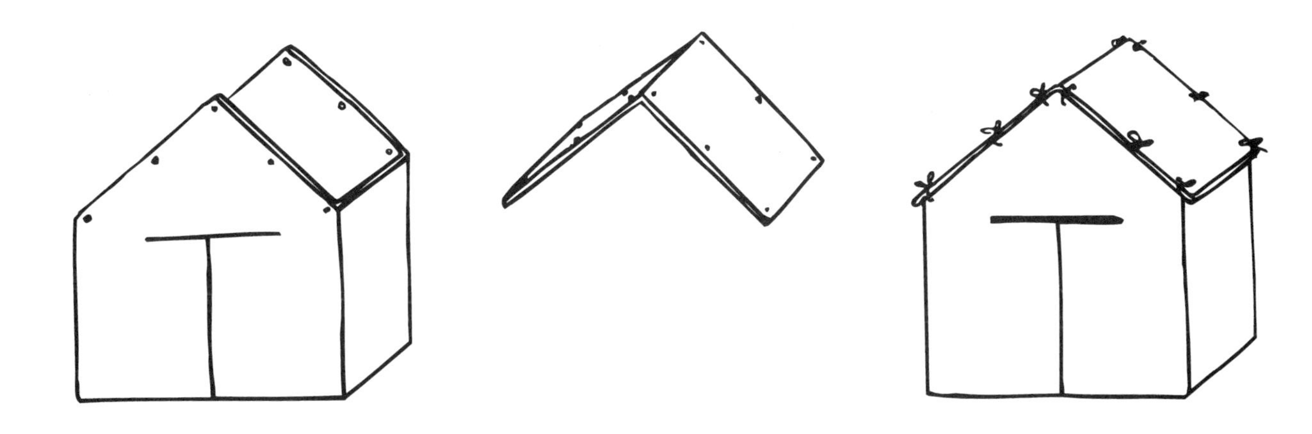

6. Have children paint the barn red and the roof brown.
7. Once the paint has dried, children can highlight the doors and the loft with white paint.

Options:

- Provide stuffed animals for farm animals.
- Let children build fences from wooden blocks.
- Cut yarn or strips of paper to simulate hay.

Book Links:

- *Big Red Barn* by Margaret Wise Brown (HarperCollins)
- *The Barn Party* by Nancy Tafuri (Greenwillow)

Tractor

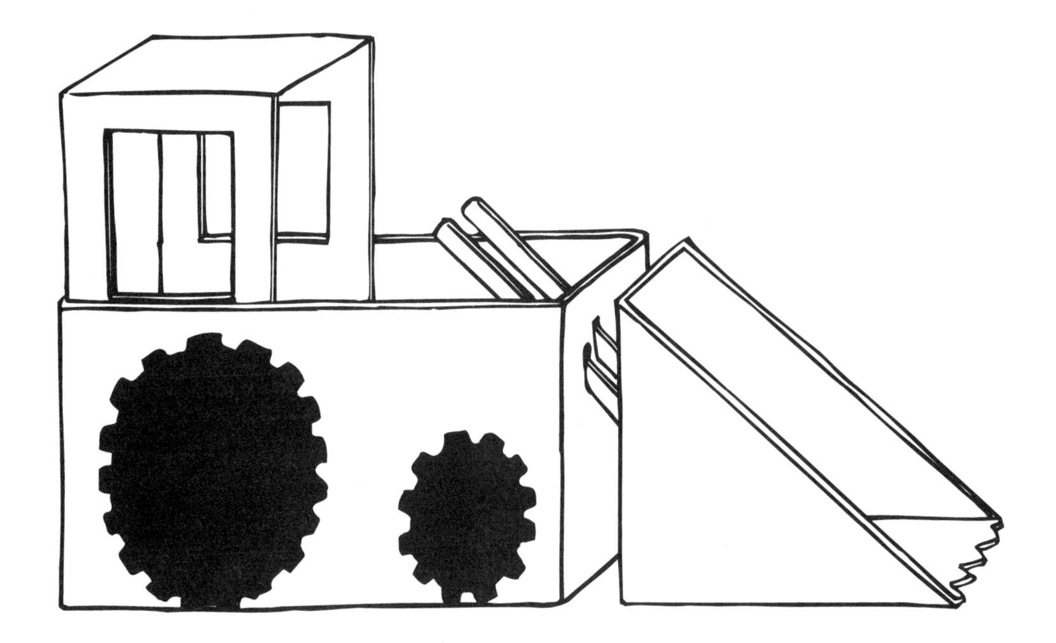

Materials:

Three cardboard boxes (large enough for a child to sit in), two wooden dowels, black construction paper, tempera paint (green, yellow, black, and white), paintbrushes or rollers, shallow tins (for paint), sharp instrument for cutting (for adult use only), scissors, glue, masking tape, black tape

Directions:

1. Remove one side of all three boxes.
2. Cut two side windows in the top half of one box.
3. Lay the second box on its side and make a diagonal cut from the corner to the middle of the bottom. (This leaves you with a triangular-shaped box.)

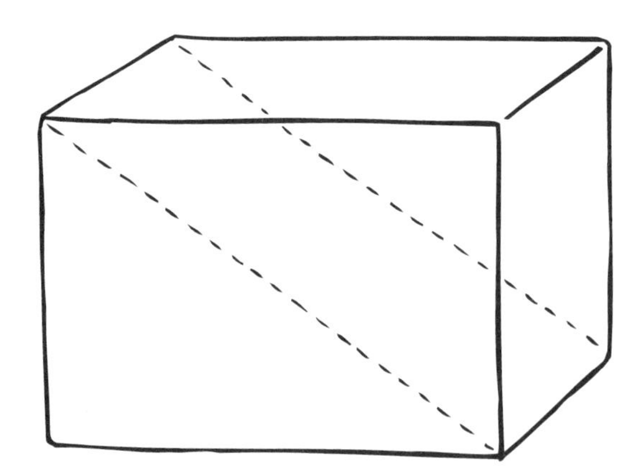

Tractor

4. Place the box with the windows upright inside the other box. The open side will be facing forward.
5. Cut two holes in the upper end of the large box.
6. Cut two holes in the middle of the triangular box and two holes in the bottom of the box.
7. Insert the dowels side-by-side through all the holes to attach the boxes. (The children can use the dowels as handles to raise the tractor's shovel.)
8. Wrap masking tape around the end of the wooden dowels to prevent slipping.
9. Have children paint the tractor green and the shovel and wooden dowels gray. Wrap the handle-ends of the dowels with black tape.
10. Children can highlight the windows with yellow paint, and cut tires from black construction paper to glue onto the tractor. (Black tempera paint can also be used.)

Options:

- Provide plastic fruit and vegetables for dramatic play.
- Let children wear plastic construction worker hats.

Book Link:

- *Tractor Factory* by Elinor Bagenal and Steve Augarde (Artists and Writers Guild Books)

Sawhorse Cow

Materials:
Cow Head Pattern (pull-out pattern), sawhorse, flat section of large cardboard box, latex glove, heavy string or yarn, tempera paint (white, black, blue, and red), sponge brushes and wide paintbrushes, shallow tins (for paint), sharp instrument for cutting (for adult use only), saw (for adult use only), hammer and nails (for adult use only), water or milk, bucket, scissors

Directions:
1. Cut sawhorse to height that allows children to sit on it safely.
2. Duplicate the cow head pattern and cut out.
3. On the flat piece of cardboard, trace the cow head pattern twice and cut out. Cut a notch in the neck to allow for the sawhorse legs. Nail the heads on both sides of the sawhorse, tying them together around the ear with yarn.
4. On a piece of cardboard, draw and cut a cow's tail. Nail the tail to the cow.
5. Have children paint the cow white.

Sawhorse Cow

6. Once the paint has dried, children can sponge paint black marks on the cow. (Or children can handprint markings.)
7. Children can paint the cow's eyes blue and the mouth red.
8. Once the cow is completely dry, make a pinhole in the tip of three fingers of the latex glove. Pour milk or water into the glove and tie the top closed.
9. Attach the glove with yarn to the sawhorse cow.
10. Children can now place the bucket beneath the cow's glove "udder" and "milk" the cow.

Option:

- Provide a small stool for children to sit on while milking.

Book Links:

- *The Cow That Went Oink* by Bernard Most (Harcourt Brace)
- *Daddy Played Music for the Cows* by Maryann Weidt and Henri Sorenson (Lothrop, Lee and Shepard)

Fruit and Vegetable Stand

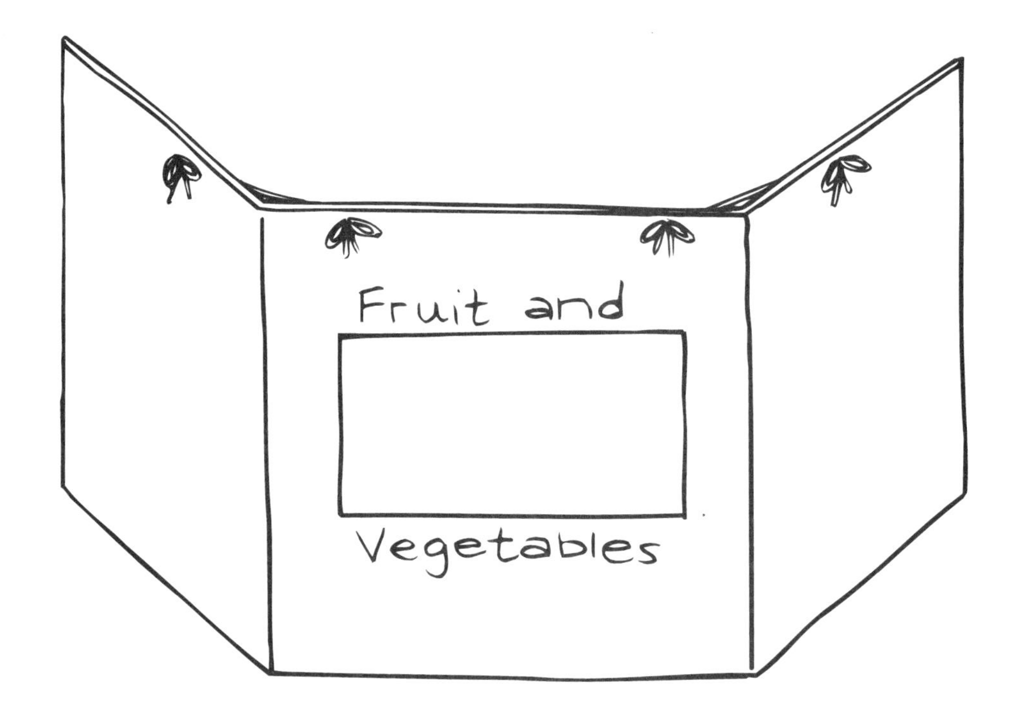

Materials:
Washer or dryer box, tempera paint, paintbrushes or rollers, scissors, shallow tins (for paint), heavy string, sharp instrument for cutting (for adult use only)

Directions:
1. Completely cut off the two ends and one side panel of the box.
2. Cut a large window in the center panel of the box.
3. Lay the fruit and vegetable stand flat and provide tempera paint for the children to use to decorate it.
4. Punch two holes in the center panel above the window and one hole in each of the side panels.
5. Reinforce the side panels to the center panel by tying the sections together with string.

Options:
- Paint the words "Fruit and Vegetable Stand" on the box.
- Make a hanging "Open" and "Closed" sign from cardboard and yarn.
- Provide a toy cash register and money, aprons, grocery bags or baskets, and plastic fruit and vegetables.

Farmer's Vest

Materials:

Vest Pocket Pattern (bottom of next page), paper grocery bag (one per child), construction paper, blue tempera paint, shallow tin (for paint), nylon soap scrunchies (these come with liquid body soap) or pieces of sponge, scraps of fabric, poster board, scissors, pencils, stapler or tape

Directions:

1. Cut each bag up the front of one side. Continue cutting a circle on the top of the bag.
2. Cut a circle on both narrow sides of the bag for the arms and a V-shape on the bottom of each panel on the front side.
3. Turn the bags inside out.

Farmer's Vest

4. Trace the vest pocket pattern onto poster board and cut out. Make several for the children to use as templates.
5. Have the children trace the pockets onto construction paper and cut out.
6. Help children staple or tape the pockets onto the vests.
7. Provide blue tempera paint and soap scrunchies or sponges for the children to use to paint their vests.
8. Once the vests are dry, let children choose scraps of fabric to stuff in their pockets as handkerchiefs.

Options:

- Provide buttons for children to glue onto their vests.
- Punch holes around the pockets and stitch the pockets with yarn before attaching to the vests.
- Provide bandannas for children to wear with their vests.

Book Links:

- *Old MacDonald* by Jessica Souhami (Orchard)
- *The Thing That Bothered Farmer Brown* by Teri Sloat (Orchard)

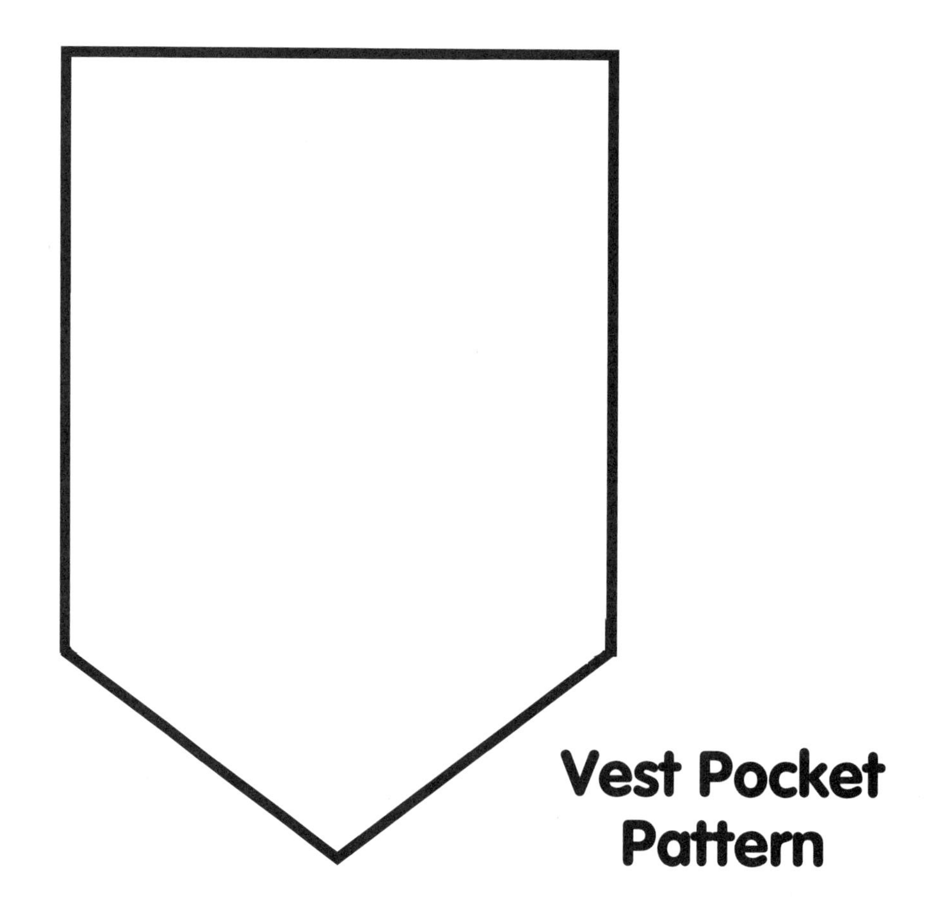

Farmer's Hat

Materials:

Hat Brim Pattern (pull-out pattern), lunch-size paper bag (one per child), poster board, construction paper (assorted colors), brown tempera paint, paintbrushes,"collage" items (beads, sequins, cut lace, buttons), glue, stapler, scissors, marker, yarn, tape, hole punch

Directions:

1. Duplicate and trace the hat brim pattern onto poster board and cut out. (Make one per child.)
2. Use a marker to divide each paper bag in half, width-wise.
3. For the hat bands, cut colored construction paper strips long enough to fit around the paper bags.
4. Provide scissors for children to use to cut their bags in half.
5. Have children open their bags. Help each child insert the open bag through the brim of the hat, then staple the two together.
6. Have children paint their hats brown.
7. While the hats are drying, children can make and decorate hat bands by gluing collage items to the strips of construction paper.
8. Once the hats and bands are dry, children can tape the bands to the hats. Help them to curl the brims of their hats.
9. Punch a hole on each side of each hat. Attach yarn for children to use to fasten their hats.

Book Link:

• *I'm a Jolly Farmer* by Julie La Come (Candlewick Press)

Pig Mask

Materials:

Ear and Snout Patterns (p. 15), dinner-size paper plate (one per child), sturdy paper, construction paper (pink and black), pink tempera paint, shallow tin (for paint), sponge pieces, Popsicle sticks (one per child), masking tape, scissors, glue, black markers

Directions:

1. Cut two holes in each plate for the eyes, and tape a Popsicle stick to the back of each plate for a handle.
2. Duplicate the ear and snout patterns onto sturdy paper and cut out. Make several for the children to use as templates.
3. Have children sponge paint their masks pink.
4. Children can trace the ears and snout patterns onto pink or black construction paper and cut out.
5. Children can use the markers to draw nostrils on their snouts.
6. Have children glue the ears and snouts onto their masks.

Book Links:

- *Mrs. Potter's Pig* by Phyllis Root (Candlewick Press)
- *Oink! Moo! How Do You Do?* by Grace Maccarone (Cartwheel Books)
- *Pigs from A to Z* by Arthur Geisert (Houghton Mifflin)

Ear and Snout Patterns

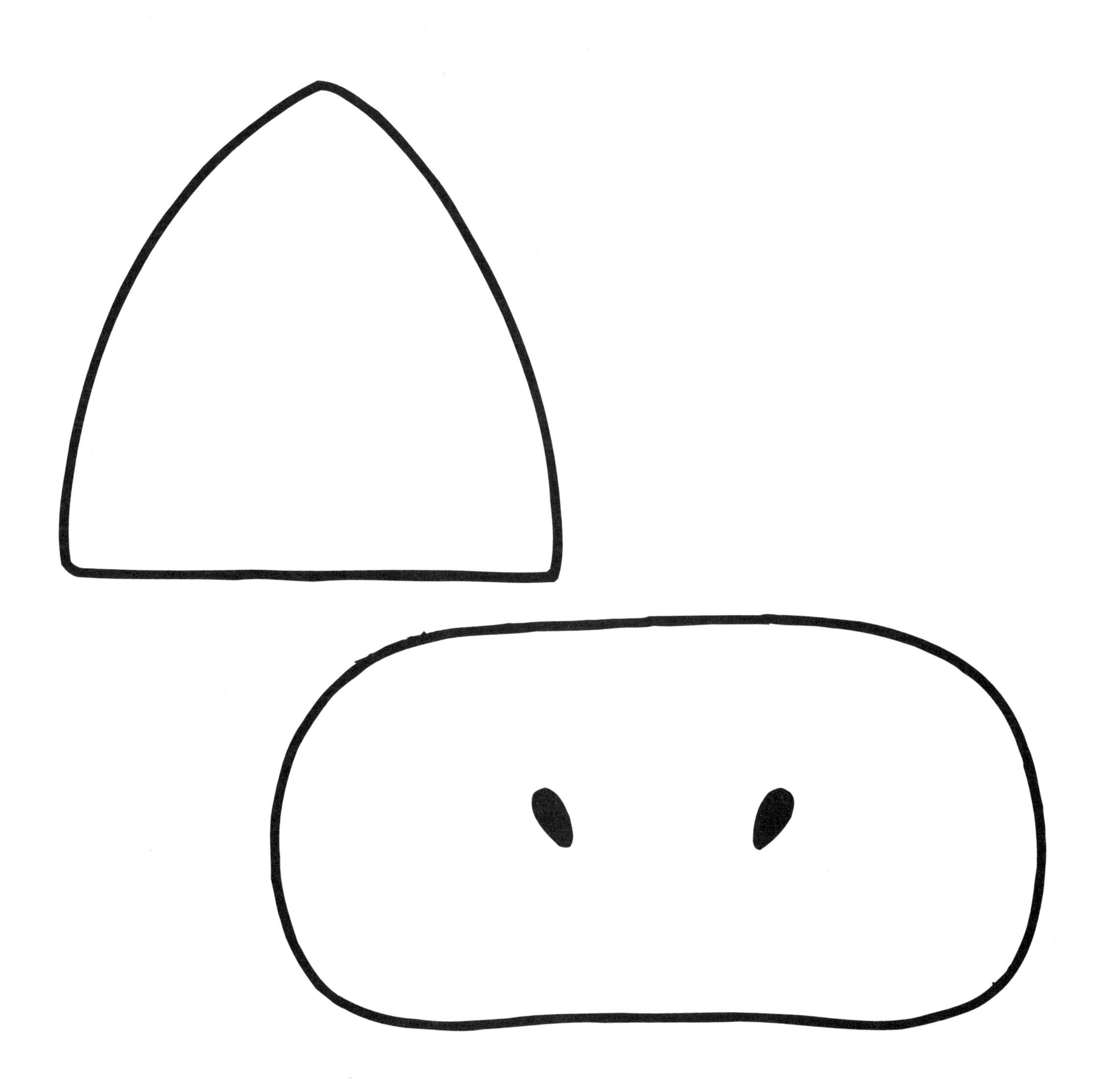

Riding Pony

Materials:

Pony Head Pattern (p. 18), construction paper (brown, black, gray, or white), newspaper, yarn, masking tape, hole punch, crayons or markers (including red), stapler, scissors

Directions:

1. Roll several long sheets of newspaper together and tape tightly. Make one per child. These will be used for the ponies' poles.
2. Trace the pony head pattern onto two pieces of construction paper and cut out. Make two pony heads per child.
3. Align both heads and punch holes around the edges, except around the ear. Draw a red circle around the last hole before the ear. (Do this for each pony head.)
4. For each child, cut several pieces of yarn (about one foot/.3 meters long), and cut one piece of yarn long enough to lace around the holes of each pony's head.
5. For easier lacing, wrap a small piece of masking tape around one end on each long piece of yarn. Tie the other end to the neck of the pony, joining the two pieces of paper together.

Riding Pony

6. Let each child lace around the front of his or her pony's head, stopping at the red circle.

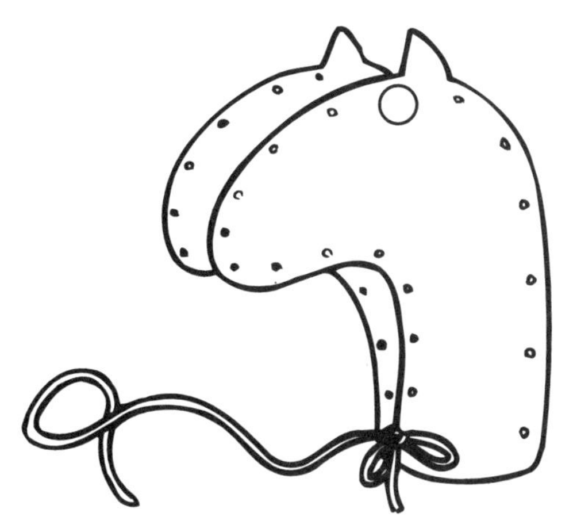

7. Children can use the other pieces of yarn to make loop knot manes. Demonstrate how to fold the yarn in half, insert the loop of yarn through the hole that is behind the pony's ear, bring the two loose ends of yarn through the loop, and pull gently. Continue for each hole to complete the mane.

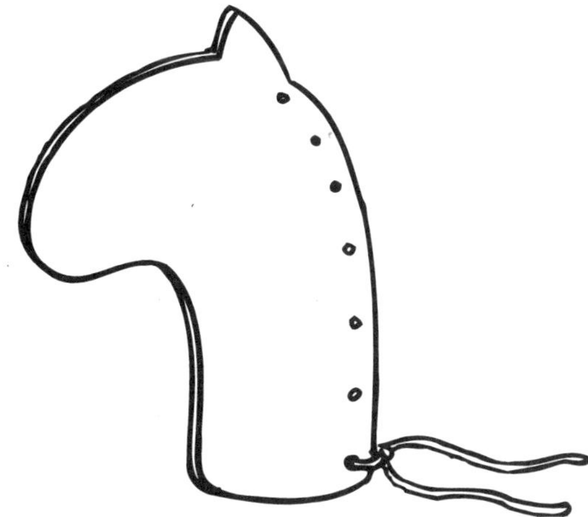

8. Provide crayons or markers for children to add details to their ponies.
9. Insert a newspaper pole into each pony's neck and staple.

Options:

- Have the children name their ponies. Write each pony's name on its neck.
- Add reins by tying two long pieces of yarn through the bottom hole of each mane.

Book Link:

- *Horses and Ponies* (Usborne Publishing)

Pony Head Pattern

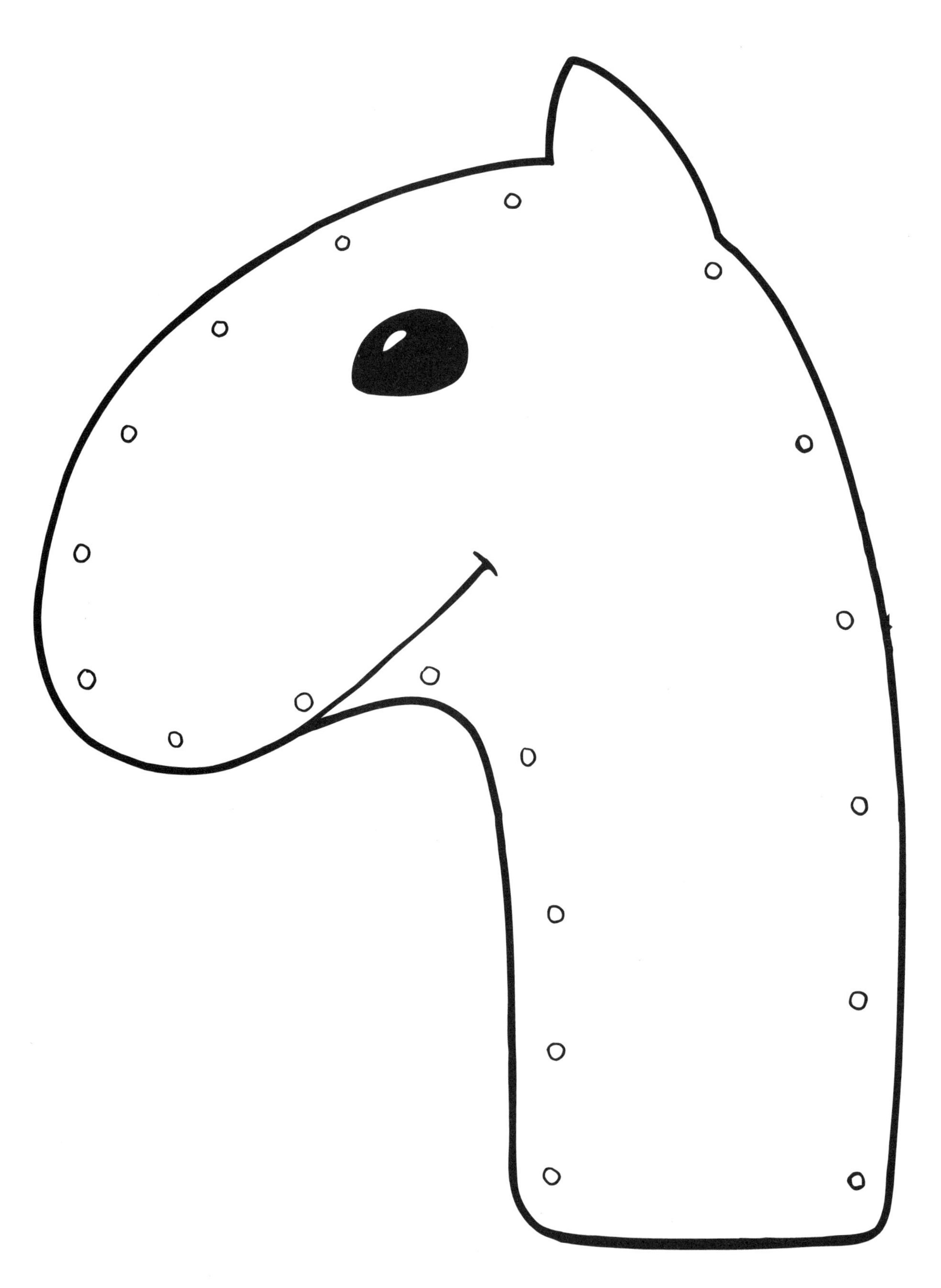

Duck Costume

Materials:

Duck Patterns (p. 21), construction paper (yellow, blue, and orange), pipe cleaners (cut in half), self-sticking dots, hole punch, scissors, tape, glue, marker

Directions:

1. Cut the yellow construction paper into three-inch (7.5 cm.) wide strips. (They should be long enough to fit around a child's head.) Make one strip for each child.
2. For each child, trace the eye pattern onto a piece of folded blue construction paper.
3. For each child, trace the webbed feet pattern onto a piece of folded orange construction paper.
4. For each child, trace the duck's bill pattern onto the fold of a piece of orange construction paper.

Duck Costume

5. Have the children cut out the eyes and feet, keeping the papers folded. (This will give them two of each.)
6. Once the children cut out the feet, place self-sticking dots on the feet as indicated on the pattern.
7. Punch a hole through each dot, and attach half of a pipe cleaner to each hole.
7. Have the children cut out the ducks' bills, keeping the papers folded. (They should not cut on the folds.)
8. Help each child glue or tape the eyes and opened duck's bill to the center of a strip of yellow construction paper. Then fit the headband to the child's head and tape the end pieces together.

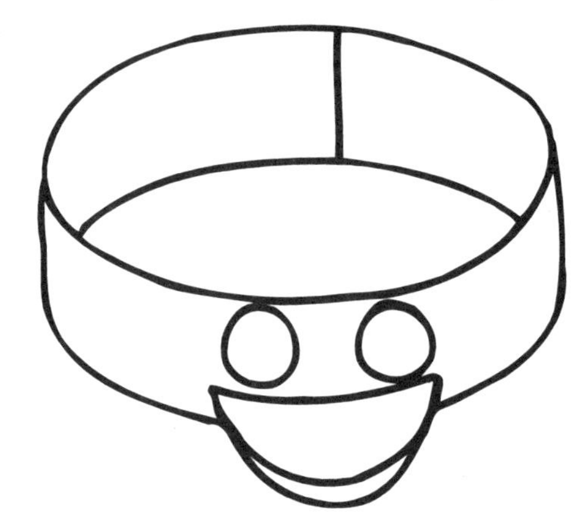

9. Help children fasten the webbed feet around their ankles by twisting the two pipe cleaners together.

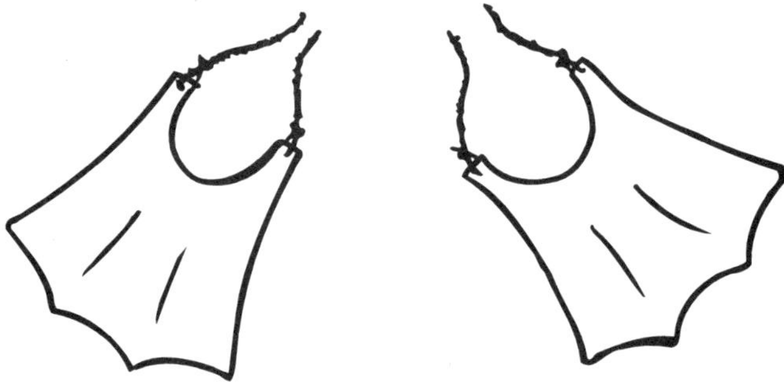

Book Link:
- *Danny's Duck* by Jane Crebbin (Candlewick Press)

Hat Brim Pattern

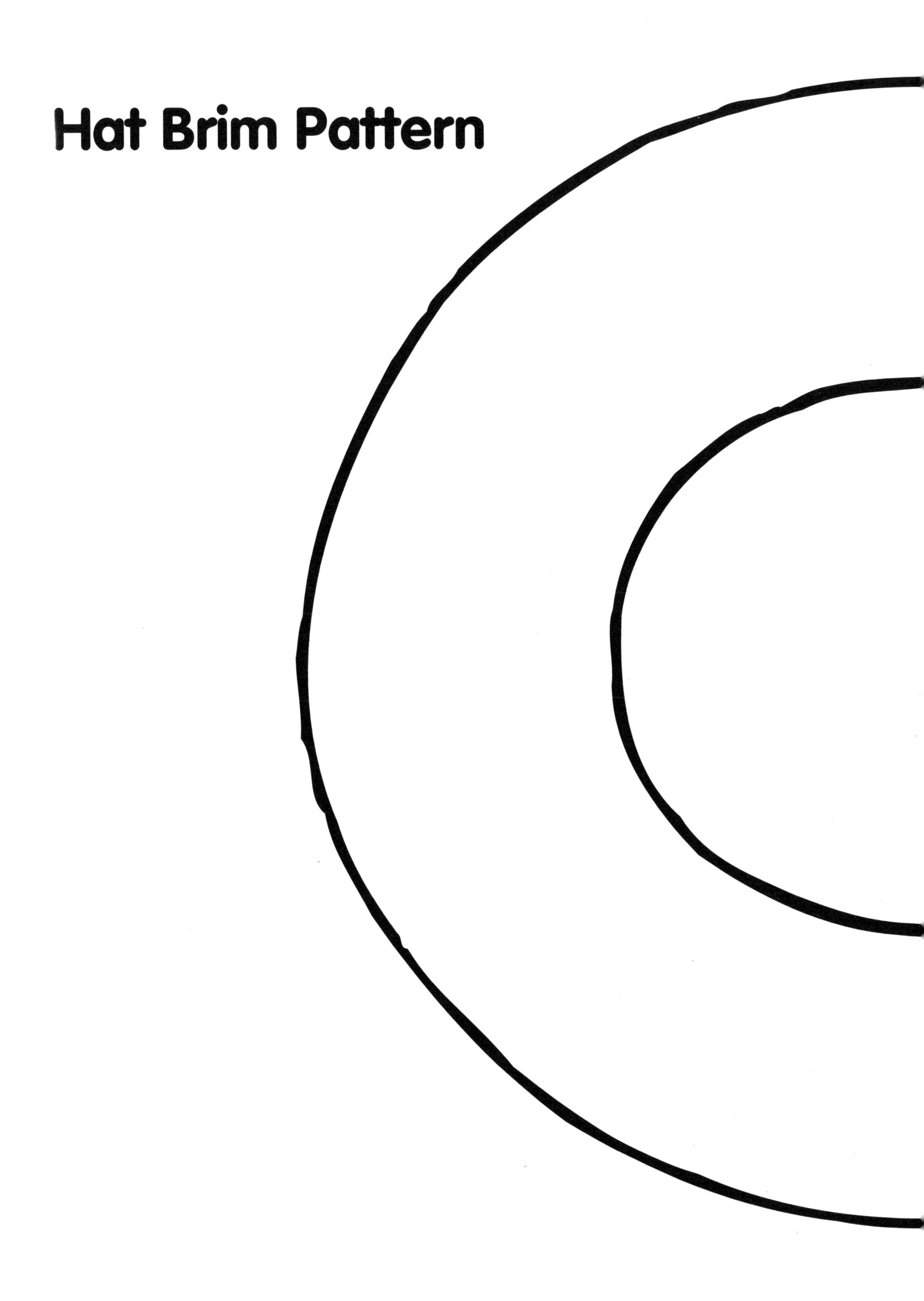

Cow Head Pattern

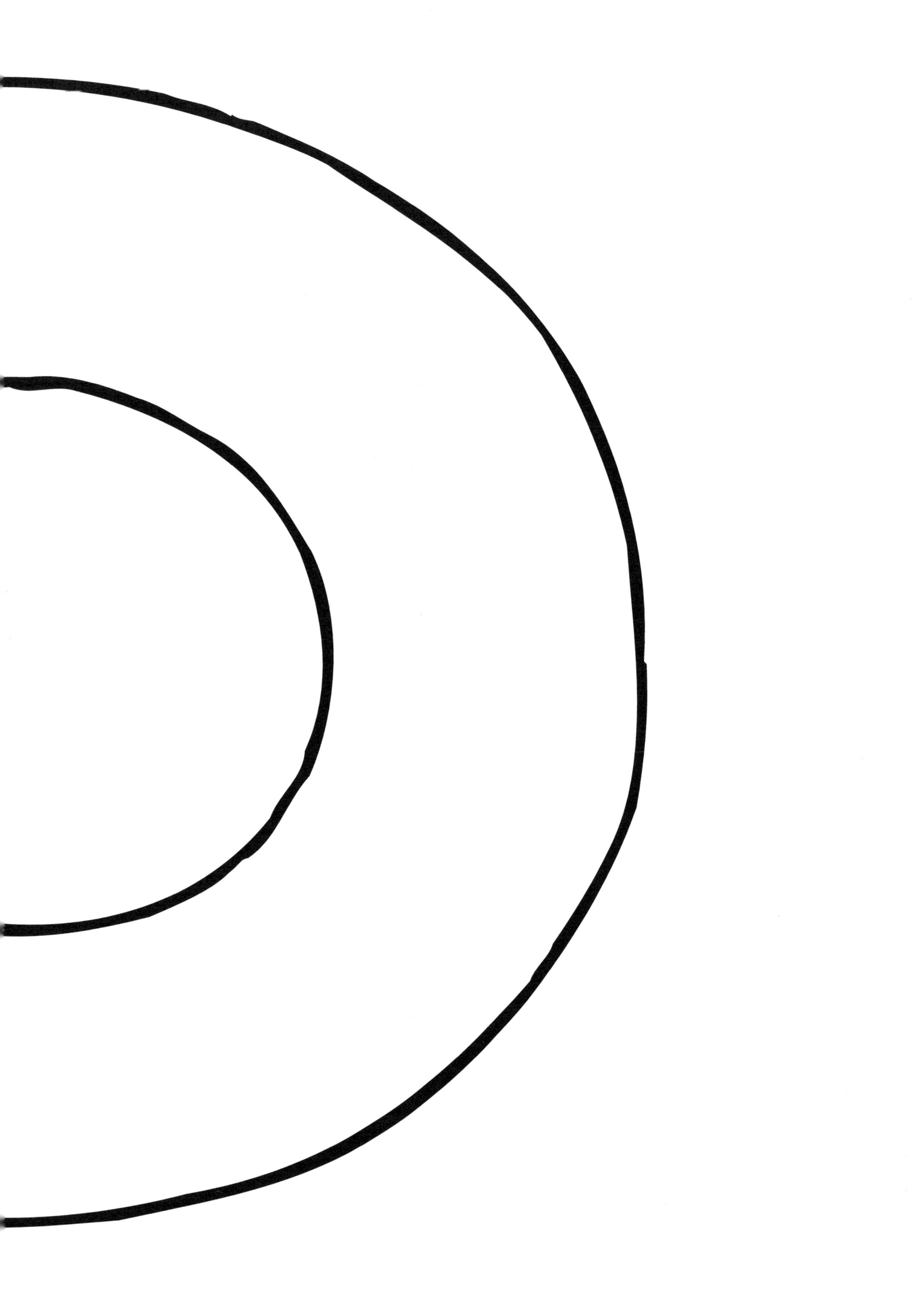

Duck Patterns

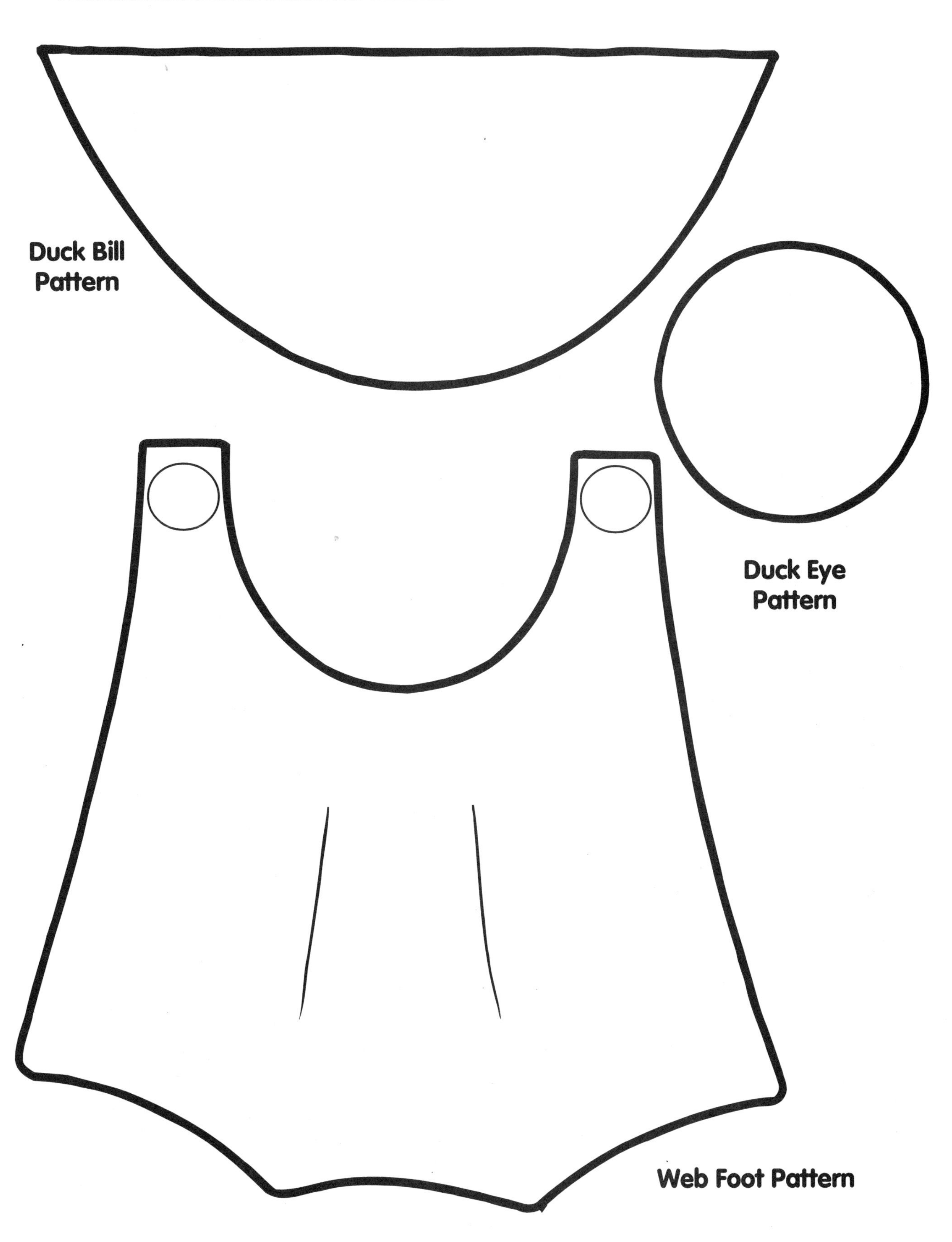

Milk Carton Barn

This activity should be used with the Fold-and-Cut Sheep (p. 28), Marshmallow Pig (p. 23), Juice Container Cow (p. 24), Clothespin Horse (p. 30), and Playdough Duck (p. 26).

Materials:
Large milk carton (clean), gift box lid, paper towel tube, glue, red construction paper scraps, tempera paint (green and brown), shallow tins (for paint), paintbrushes, markers, stapler, sharp instrument for cutting (for adult use only)

Directions:
1. Staple the top of the milk carton closed and cut out a T-shaped door on one side.
2. Have children glue ripped and torn red paper scraps to the carton.
3. Have children paint the paper towel tube brown.
4. Once the tube and the barn have dried, glue the paper towel tube to the side of the barn for the silo.
5. Children can add details to the barn with markers.
6. Let children paint the gift box lid green for the pasture.
7. Once the paint has dried, place the barn in the pasture.

Options:
- Use brown construction paper for the barn roof, rice to fill the silo, strips of green paper for grass, and Popsicle sticks for a fence.

Book Link:
- *Barn Dance* by Bill Martin Jr. and John Archambault (Holt)

Marshmallow Pig

Materials:

Marshmallows (regular and miniature), toothpicks, pink pipe cleaners, pink tempera paint, small paintbrushes, small cups (for paint), scissors

Directions:

1. Cut small sections of pipe cleaners for ears and tails.
2. Give each child a large marshmallow for the pigs' bodies.
3. Have children insert toothpicks and miniature marshmallows into the bottom of the large marshmallow. (These will be the pigs' legs and feet.)
4. Children can insert toothpicks and miniature marshmallows into the front of the large marshmallows for the pigs' snouts and eyes, as shown.
5. Children can insert pink pipe cleaner pieces for the pigs' ears and curly tails.
6. Provide pink tempera paint for the children to use to paint their finished pigs.

Book Links:

- *I Like Me!* by Nancy Carlson (Viking)
- *Pigs Aplenty, Pigs Galore* by David McPhail (Dutton)
- *The Pig in the Pond* by Martin Waddell (Candlewick Press)

Juice Container Cow

Materials:
Face and Horn Patterns (bottom of next page), plastic juice container with lid (individual-size; one per child), golf tees (five per child), black and white construction paper, glue sticks, scissors, markers

Directions:
1. Rinse out the juice containers and replace the lids.
2. Using a golf tee, punch four holes in the four corners of one side of each container. (This is where the legs will be inserted.)
3. Punch one hole in the end of each container for the tail.
4. Trace the face pattern onto white construction paper and cut out. Make one for each child. (Or duplicate a face pattern for each child to cut out.)
5. Trace the horn pattern onto black construction paper and cut out. Make one for each child. (Or duplicate a set of horns for each child to cut out.)

Juice Container Cow

6. Cut small circles from black and white construction paper for spots on the cow.
7. Provide golf tees for children to insert in the holes for each cow's legs and tail.
8. Have children use markers to color the cows' faces.
9. Each child can glue a face and horns to a lid, making sure the lid is on tight.
10. Have children glue the construction paper spots to the cows.

Book Links:
- *Farm Animals* by Rowena Holland (The Bookwright Press)
- *Farm Babies* by Russell Friedman (Holiday House)

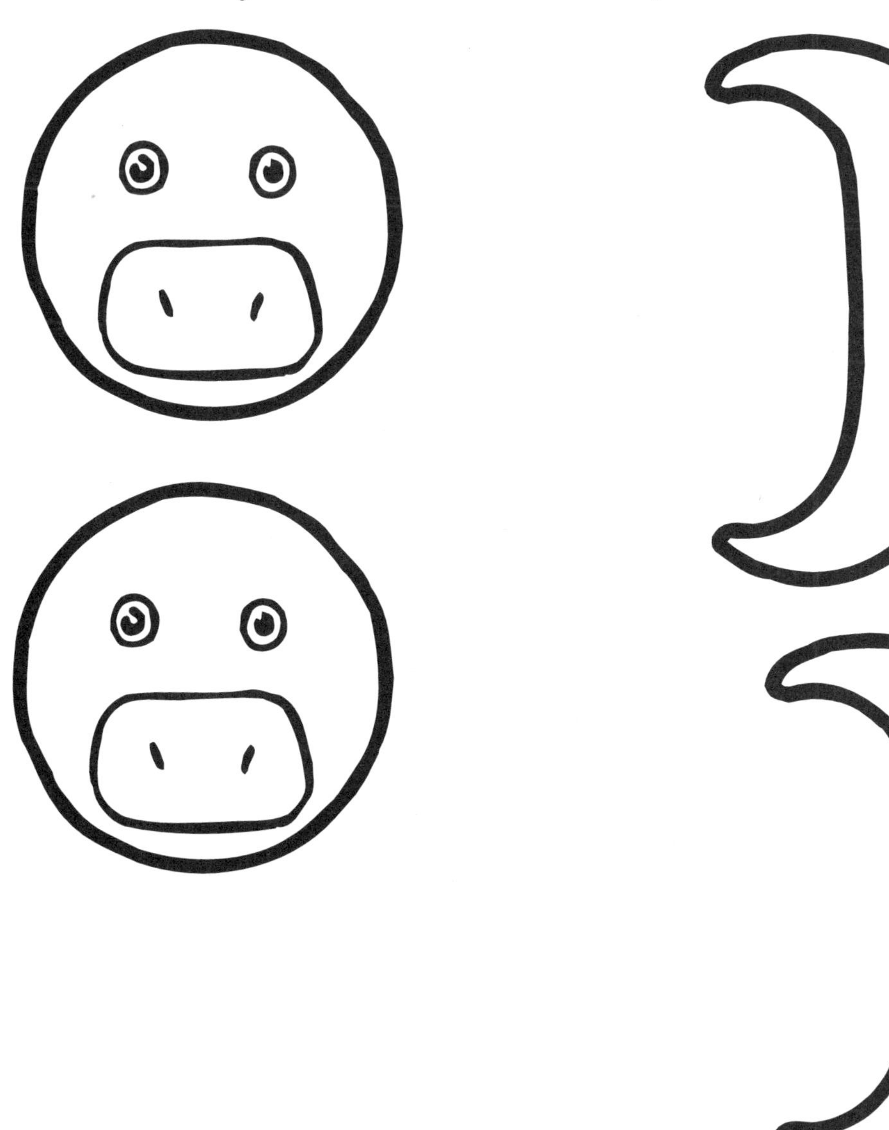

Playdough Duck

Materials:

Head and Feet Patterns (p. 27), playdough ingredients (see recipe below), sturdy paper or old file folders, feathers (orange, white, or yellow), crayons, scissors

Directions:

1. Make the playdough with the children, using caution when adding the boiling water.
2. Trace the head and feet patterns onto the file folders and cut out. Make one set per child. (Or duplicate the head and feet patterns for children to cut out and glue to sturdy paper.)
3. Have children color the ducks' heads and feet with crayons.
4. Once the playdough has cooled, give each child a small ball of playdough to use for the duck's body.
5. Have the children insert the head and feet patterns into their balls of playdough, and then insert the feathers.

Playdough Recipe

4 cups (1 kilogram) flour
2 cups (.5 kilograms) salt
8 tsp. (40 grams) cream of tartar
10 tsp. (50 ml) liquid vegetable oil
4 cups (1 liter) boiling water
food coloring (desired color)

Directions:

1. Combine the first four ingredients in a large bowl.
2. Add food coloring to the boiling water.
3. Pour the water into the dry ingredients and mix.
4. Remove the dough from the bowl and knead on a floured surface.

Book Link:

• *Farmer Duck* by Martin Waddell and Helen Oxenbury (Candlewick Press)

Head and Feet Patterns

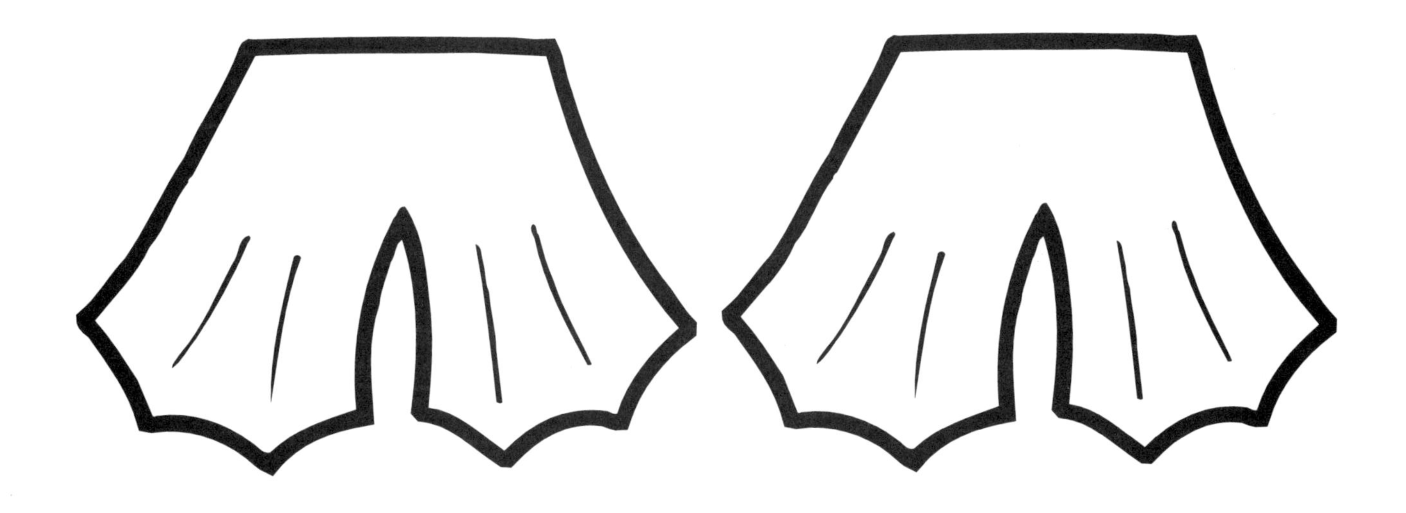

Fold-and-Cut Sheep

Materials:

Sheep Patterns (p. 29), white construction paper, white ribbed ribbon, markers, scissors, glue stick

Directions:

1. Use scissors to curl long strips of ribbon. Cut the strips into small pieces to be used for the wool.
2. For each child, trace the sheep head pattern onto white construction paper and cut out. (Or duplicate a sheep head pattern for each child and cut out.)
3. For each child, trace the sheep body pattern onto the fold of a piece of folded white construction paper. Cut a notch for the head at the fold.
4. Keeping the paper folded, have children cut out the sheep. They should *not* cut on the fold.
5. Provide markers for children to use to color the faces.
6. Show children how to insert the sheep's head into the notch of the body. (Or glue or staple each head to its body.)
7. Children can glue the curled ribbon to their sheep for wool.

Book Links:

- *Sherman the Sheep* by Kevin Kiser (Macmillan)
- *Spring Fleece: A Day of Sheepshearing* by Catherine Paladino (Little, Brown)

Sheep Patterns

Clothespin Horse

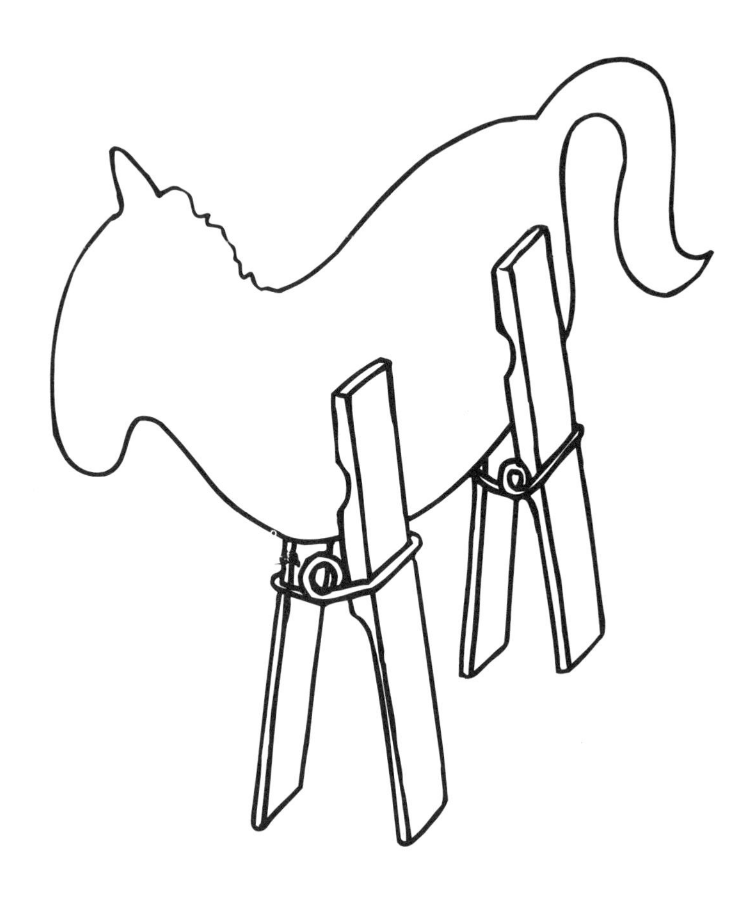

Materials:
Horse Pattern (p. 31), old file folders, clothespins (two per child), tempera paint (brown, black, or white), shallow tins (for paint), cotton swabs, scissors, marker

Directions:
1. For each child, trace the horse pattern onto a file folder and cut out.
2. Give each child two clothespins to attach to his or her horse for legs.
3. Provide tempera paint and cotton swabs for children to use to paint their horses.

Options:
• Remove the tail from the horse pattern and punch a hole at the back of each horse, instead. Children can loop a length of yarn through the holes to make tails. (Refer to "Riding Pony" pp. 16-17.)

Horse Pattern

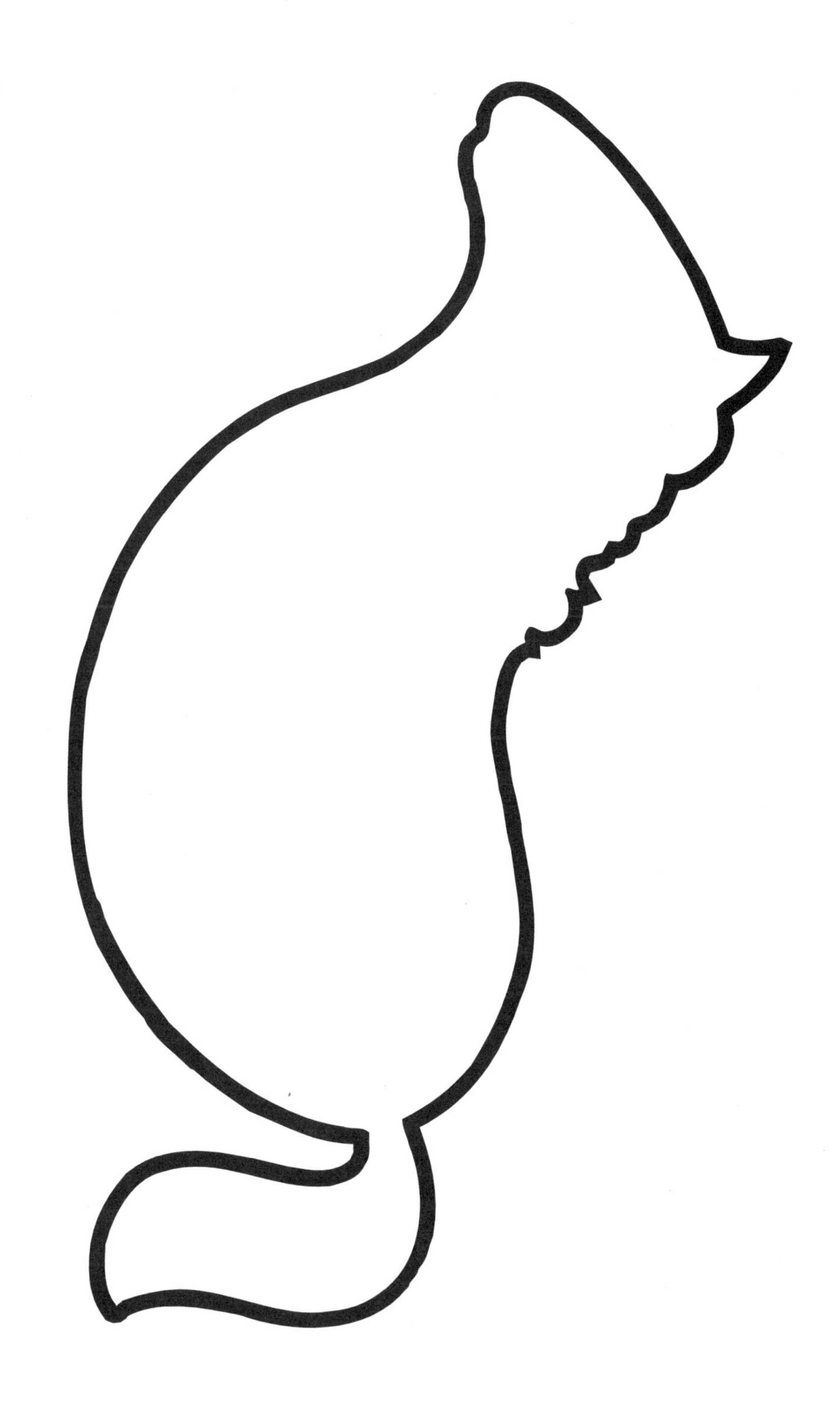

Seed Growth Mural

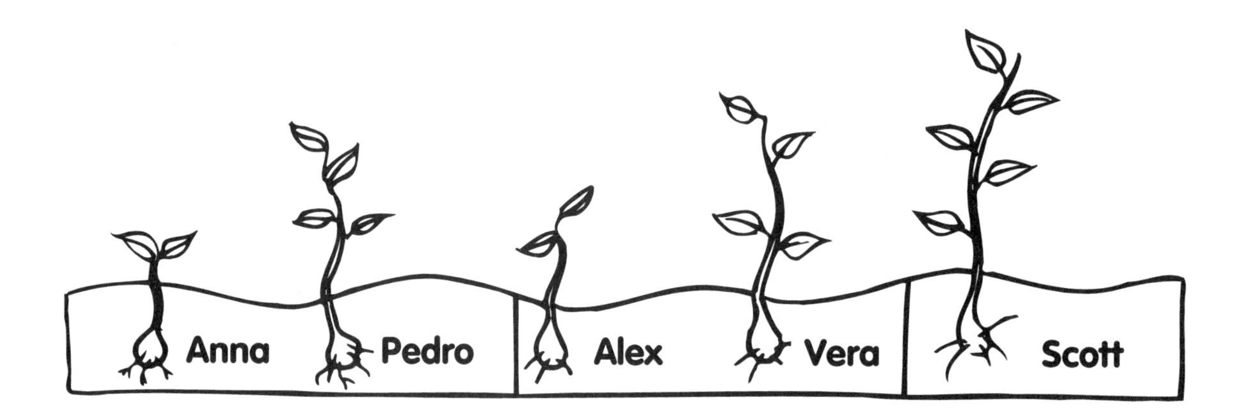

Materials:
Seed and Leaf Patterns (p. 34), construction paper (brown, white, and green), sturdy paper, beige or white yarn, glue or glue sticks, scissors, markers

Directions:
1. Cut long strips of brown construction paper approximately eight inches (20 cm.) wide. (Tape several sections together, if needed.)
2. Trace the seed and leaf patterns onto the sturdy paper and cut out. Make several for children to share as tracing templates. (Or duplicate the seed and leaf patterns onto colored paper for children to cut out.)
3. Cut short pieces of yarn. (These will be the roots.)
4. On Day One, discuss the growth sequence of a seed with the children. Provide seed patterns and white construction paper for the children to use to trace the seeds and cut out. Children then glue the seeds onto brown construction paper. (Write the children's names next to their seeds.)

Seed Growth Mural

5. On Day Two, discuss why the seeds grow roots first. (Refer to book links if needed.) Provide short pieces of yarn for the children to glue to the bottoms of their seeds.
6. On Day Three, discuss the growth of a seedling. Provide leaf pattern templates and green construction paper for the children to use to trace the leaf and then cut out. Children should also cut strips of green construction paper for the stems. Have the children glue the stems and leaves above their seeds.

Options:
- Hang paper clouds, raindrops, and a sun above the mural.
- Let children drink juice or milk through straws to experience how a seed absorbs nourishment through its roots.

Book Links:
- *The Big Green Bean* by Marcia Wiesbauer (Silver Press)
- *Growing Vegetable Soup* by Lois Ehlert (Voyager Books)
- *Vegetable Garden* by Douglas Florian (Voyager Books)

Video Links:
- *The Magic School Bus Goes to Seed* (Scholastic)

Note:
This activity can be used with the Seed Bottles (p. 35) and Growth Cards (p. 36).

Seed and Leaf Patterns

Seed Bottles

Materials:

Clear two-liter soft drink bottle (bottom portion should also be clear), potting soil, lima beans and popcorn kernels, spray bottles, plastic spoons, scissors, tape

Directions:

1. Rinse the soft drink bottle. Approximately three inches (7.5 cm.) from the bottom, cut around the bottle. Leave it attached to one side by an inch (2.5 cm).
2. Let children fill the bottom portion of the bottle with the soil.
3. Children can use the spoons to dig small holes for the seeds. They can plant the seeds and water them with the spray bottles.
4. Tape the bottle back together. Put the lid on tight and place the bottle in sunlight.
5. On a daily basis, have children remove the lid and spray water on the seeds. (Replace the lid each time and return the bottle to the sunny area.) Stop watering after one week. (Condensation will provide enough moisture for the seeds to grow.)
6. Children should observe the growth of the seeds (through the clear bottom of the bottle).
7. After the plants have grown substantially, remove the lid.

Note:

These directions are for one bottle. You can make individual bottles for each child, or several for children to share.

Growth Cards

Materials:
Growth Cards (bottom of the page), crayons, scissors, glue sticks, plain paper (one sheet per child), pen

Directions:
1. Duplicate one copy of the growth cards per child.
2. Have children color the cards and cut them apart.
3. Children should glue the pictures in the correct order on plain paper, and dictate a story about the pictures. Transcribe the story beneath the pictures.

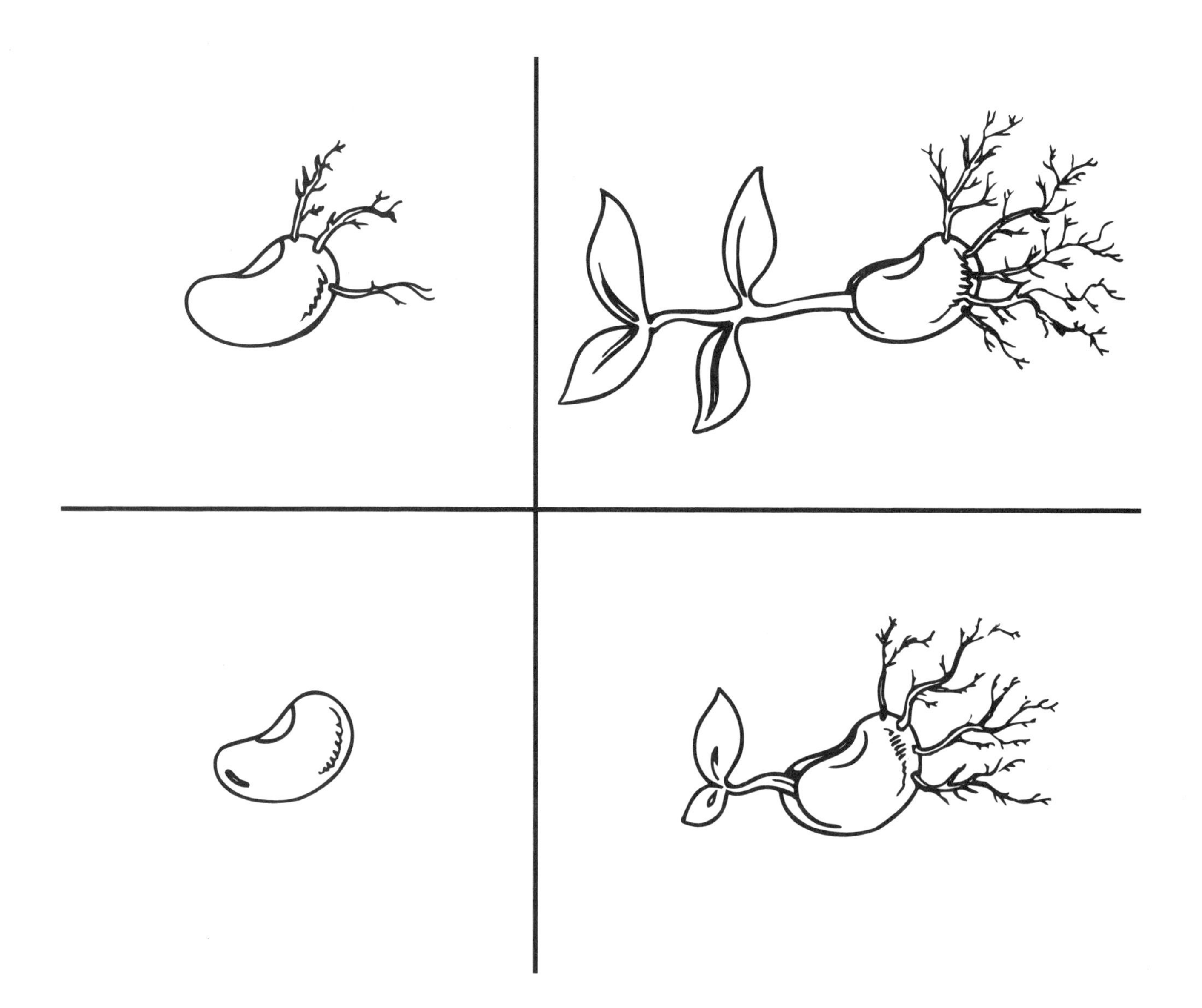

Number Match

Materials:

Number Match Pattern (p. 38), crayons or markers, yellow self-sticking dots

Directions:

1. Duplicate a copy of the number match pattern for each child.
2. Write the numbers 1 to 10 on the self-sticking dots. Make a set of dots for each child.
3. Have the children match the numbered dots to the corresponding numbers on the pattern.
4. Provide crayons or markers for children to use to color in their pictures.

Options:

- If self-sticking dots are not available, cut circles from yellow construction paper and number these. Children can glue the circles to the patterns.
- Write the numbers out of order on the yellow dots.
- For more advanced children, white-out the numbers on the pattern before duplicating it. Have children write in the numbers in order in the circles.

Number Match Pattern

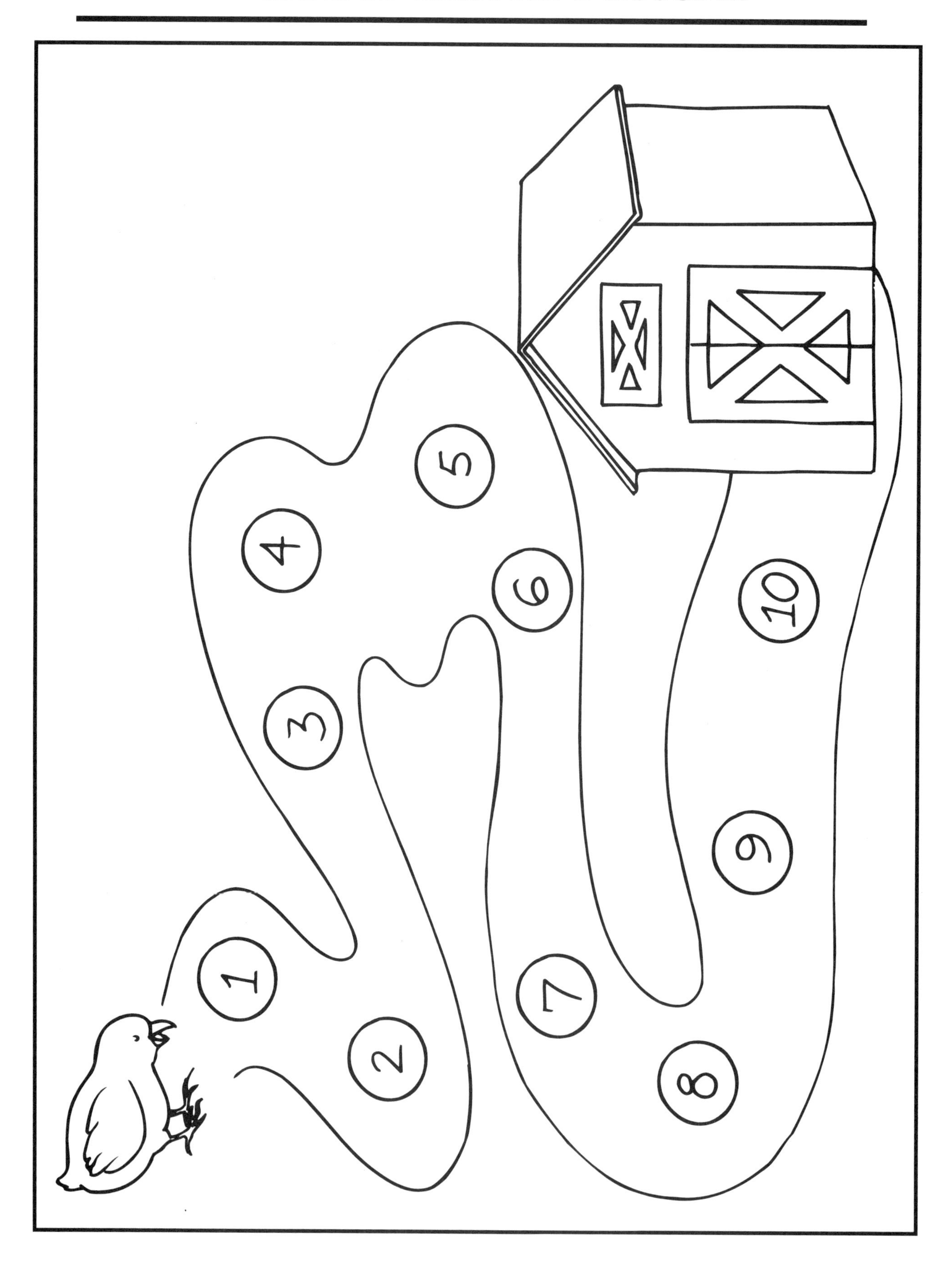

Memory Match

Materials:

Farm Life Patterns (p. 40), colored markers, scissors, clear Contac paper

Directions:

1. Duplicate the farm life patterns twice, color, cut apart, cover with Contac paper or laminate, and cut out again. (Leave a thin laminate border around each pattern to help prevent peeling.)
2. Shuffle the cards and spread them face down on a table.
3. Demonstrate how to play the game. The object is to match the farm life patterns by turning the cards over two at a time. If a match is made, the cards remain face up and the child takes another turn. If a match is not made, the cards are turned over and the next child takes a turn. The game continues until all cards are face up.

Option:

- Introduce the game by leaving the shuffled cards face up and having the children simply match the animals together.

Book Link:

- *The Rooster Who Lost His Crow* by Wendy Cheyette Lewison (Dial)

Farm Life Patterns